Never Trust A Science Teacher

Nurturing Children's Reasoning
The Essence of Thoughtful Citizenship

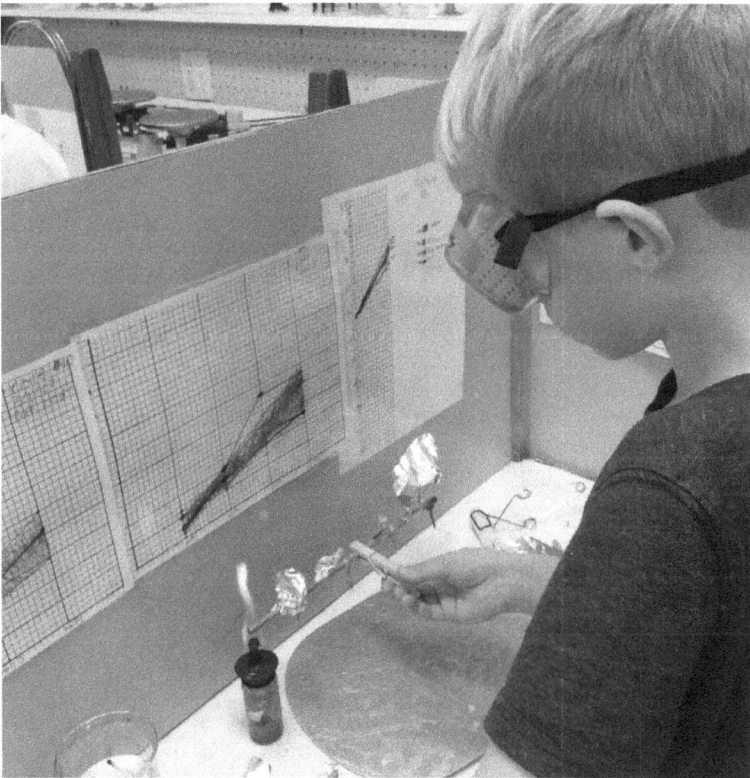

Meredith Olson, Ph.D.

http://www.DocOsBooks.com

Published by
SAN 299-2701
Glenhaven Publishing
4262 NE 125th Street
Seattle,Washington 98125

ISBN 978-0-9984627-2-1 (paperback)

Table Of Contents

1. Never Trust A Science Teacher

Never trust a science teacher. That is an odd thing to say. Do you not trust your teacher? Your mentor? Your authority figure? Don't you believe the teacher is trying to help you? To do something good for you? Don't you trust the teacher with your well being? Your development? Your life?

Never trust a science teacher. The sentence has two ideas. Should you not trust the teacher or should you not trust the science? What is it about science that is not to be trusted? Certainly our populace today is leery, suspicious, wary of science. Why is that? We hear on the news that *97% of scientists believe (something)*. Shouldn't that make us trust them?

Actually, consensus isn't what science is about. Science is about doubt. Scientist don't "believe." They predict. They look at the data. And doubt. And check. And hold theories tentatively – knowing that more data will cause them to reconsider.

Scientists are a collection of individuals. They hold onto their individuality passionately. They hold their sense of self separate and distinct from any theory. They do not allow themselves to be defined by a theory they support. To do so means that any attack on the theory is an attack on the person, on the self-esteem of the person. This places self-preservation in conflict with a considered assessment of real world data. Scientists are not unique. Anyone can become a scientist who chooses to engage in the scientific method. They are people spread throughout society. We personally know scientists and, from time to time perhaps, by our conduct we are scientists.

Scientists are not politicians. They are not salesmen. Politicians, salesmen and news reporters listen to what scientists say and then report certainties. Scientists almost never say they are certain.

Everyday citizens are confused. They hear different stories from different folks. How can the news be certain when scientists are not? How can we live with uncertainty? It is a strange contrast. Science is very powerful. It has strong forces driving it forward. It is not passive nor gingerly moving forward. It acknowledges ignorance and emphasizes doubt yet that is the way it makes progress. Because there is doubt, we look for new variables, new directions to explore. We look for new things to test. We don't know what is true, exactly, so we keep searching.

A major problem is with how science is taught. If students, future citizens, are taught that science is a collection of "facts" then they come to believe that science knows what it is talking about. If schools could somehow teach the "process" of science, future citizens might come to understand the uncertain nature of this endeavor called science. How do we do that? How do we prepare students for thoughtful citizenship?

First and foremost, science is individual. Every scientist reserves the right and responsibility to examine the data and draw their own conclusions. How do we do that in a classroom where we teach groups of students? Students all do the same experiment. They all report data on the same topic. Where does the individuality come in?

If you can't trust a science teacher, what can you trust?

2. Individuality

What is unique about citizenship in our country? What is asked of our citizens? The voting public. There are so many of us that we can't all speak in public debate as was done in Greek city states. Even so, we can vote. We can vote thoughtfully. To do that we are each given the challenge and opportunity to consider issues. Not "group – think." Not "pressure groups." But the opportunity to cast a ballot. The opportunity to contemplate an issue and express a personal opinion.

Society believes in schools. Our government believes in schools. Our taxes pay for schools because we believe that education is necessary to make thoughtful citizens. What kind of education do future citizens need? What is education anyway? What does it mean to be educated as opposed to being simply trained? In order to enter a skilled profession such as finance, medicine, law, or any endeavor of economic success, a person must read. Not just read, but study. Literacy implies analysis. Thinking in an analytical way. Not just memorization but actually understanding the implications of what is written. Reading and rereading. Extensive debates and discussions promote the ability to develop an argument. A rational analysis. These modes of thinking allow the development of legal codes. Regulations underlie organized society. Our citizenry is asked to explore regulations and cast their vote. To express an opinion about the regulations of the society in which they participate. Those opinions, hopefully, are based on serious research, study of the available data, and analysis of what others have written. So, we send children to school to enable them to develop the ability to read and the desire to reason.

What people need is a world-view by which to evaluate any legislation. They need criteria. A set of logical tests to see if the proposed legislation fits our world-view. This implies a great deal of effort to develop a sense of who we are and how we function as a society. What we want is for citizens to have a healthy skepticism of anything presented as "certain.' How is this developed in people? How do we provide a way to respond to news reports claiming that, *"Scientists are certain that..."?*

Schools, as we know them, teach lessons in Language Arts and Mathematics. But why do they teach science? What is the importance of science education? Jacob Bronowski, in his book *Science and Human Values,* discusses important values, such as a search for hidden likeness, the habit of truth, independence of rigorous thought, and respect for human dignity.

I try to promote these attitudes in my classroom. Student comments are valued in a respectful and intellectually safe way. I ask questions they have not previously considered in hopes of setting them up to get surprising results. Predictions are often wrong. Interestingly wrong. Students' ideas are respected and appreciated even when they are wrong. Having shown themselves that they have drawn reasonable sounding but wrong conclusions multiple times, students eventually acquire a cautious approach or at least keep an open mind to other possibilities. By testing theories with daily laboratory activity and debating implications of their results, students revise their theories and grow in their understanding.

Bronowski advocates the habit of truth. There are two ways to look for truth in science activity. One is to have a verbal representation of a concept (as when told by a textbook) and to examine physical objects to see if they conform to our expectations. (Example: When do trucks tip over? Do you know the simple rule?) A second way is to collect data and arrange them in ways to reveal underlying regularities.

Finding truth is tricky. It is easy to quote authority. If the information is wrong it is the authority's fault, not yours. This top heavy style of teaching requires you to be sure your work agrees with authority. Usually a child whose lab report does not agree with the textbook is assigned a low score. The child comes to believe that science does not represent the world we see – so we had better memorize the "right" answers in order to get high scores on exams. A child learns that the data he generates, his personal data points are not valued. Furthermore, students find no compelling reason to tell the truth! If they alter their lab results to conform to the text, they are rewarded with high scores. We are teaching students the value of fabricating data!

We do it differently. An amazingly different value system is derived by each lab group reporting their data points to the class so each student can create a graph showing all the class data. This shows public respect for everyone's data and allows each student to search for patterns and recognize outliers. Most importantly, each student then estimates a confidence interval to personally predict what will happen in the next experiment. Each lab feels an obligation to tell the truth as they record it – and even to indicate the types of difficulty they encountered with their equipment. ("My balance seemed rusty – it got stuck easily.") I have been continually amazed at the usefulness of this method. Students typically obtain results which distribute nicely around the scientifically accepted regression line. When asked to predict their best guess of the next data point (extrapolate), students quickly agree on some point near the linear trend. So long as their personal data point is recognized as *truthful and of value* to the development of the overall data trend – or off the trend but of value in generating a testable question – students perceive the drive for truth at the center of science. Outlier data points are celebrated for their truthfulness and often generate a great deal of learning for the entire class. Additionally, this truthfulness is to be tested in action and new theories can be devised without loss of respect for previously held concepts. You don't need to rely on authority. You can figure it out for yourself.

In *Science and Human Values* (page 60), Bronowski says:
"The values of science derive neither from the virtues of its members, nor from the finger-wagging codes of conduct by which every profession reminds itself to be good. They have grown out of the practice of science, because they are the inescapable conditions for its practice." ... "Science is the creation of concepts and their exploration in the facts. It has no other test of the concepts than its empirical truth to fact. Truth is the drive at the center of science; it must have the habit of truth, not as a dogma but as a process."

Why does our populace resent, fear, misunderstand the process of science? A child's thinking is programmed at a very young age. They benefit from repeated opportunities to see their data reveal new understanding. People become comfortable with the tentative nature of science data when they see it work for them.

Our society relies on individuality. It provides each child's identity and willingness to work. Speaking in debate is not selfishness. It is personalization of the idea under consideration. Students speak to share and engage with others with an energetic, cooperative intent as the idea is developed. The discussion makes us conscious that we are "becoming." That ideas we hold personally are in a state of flux. Thoughts are enlarging as more opinions are brought out. Students' minds are active, combative, progressing, growing. Developing individual points of view. In agreement with some but responsibly separate from the conclusions of the class as a whole.

The development of classroom individuality is a long process designed to develop and reinforce the sense of individual conscience. The ability to stand for what you believe. But to realize that may change. The intention is not to replace one authority with another. The students' cooperative intent comes from the environment set by the teacher, without which you get "Lord of the Flies." Rather, class debate is designed to develop appropriate skepticism and encourage the expression of individual thought.

In fact, classes structured for inquiry discourse do more than champion individual rational argumentation. They subtly and actively impose the value system on others. While not demanding, everyone is encouraged to contribute to debate. Most often, even the quietist students in class are eventually drawn into sharing their own personal thinking. Personal expression, personal liberty, personal recognition leads to personal striving. Members of the entire class strive to comprehend the discussion. In doing so, everyone's understanding increases. Everyone's "grades" improve.

From individuality comes originality. As students strive to express their understandings new connections occur to them. Creative connections between ideas occur to some students and provide examples of the possibility of creative thought processes to others. Free discussion "levels the playing field" in class. Adolescent energies are largely focused on defining themselves in a group. Are they popular? Athletic enough? No one focuses on who is more popular or more wealthy in our class discussions. Just who makes more connections. Who is more interesting.

Individuality permeates the world-view of the developing citizens. Good students no longer limit their focus to duty, deference and adherence to convention. The classroom dynamic gives them permission to explore their personal imagination to think of divergent contraptions to make and possibly divergent ways of creating professional careers for themselves. Today this type of citizen designs new products and new ways of doing business, often with amazing economic success. A major theme in today's economic scene is the personalization of business and the success it brings.

In my experience, classrooms run this way do not erode a sense of community. Authoritarian processes are not required to provide cohesion within a group. Individuality ties us together with a different sort of bond. A bond of mutuality, respect and

cooperation which tolerates differences of opinion. We don't all
have to agree on the cause of global warming but we do agree on
the desirability of presenting all the data and points of view we can
find. It requires more effort to consider the arguments than to
simply chant, "Save the Earth." Rather than being intellectual
strangers to our classmates, we become well acquainted with their
thinking as we listen to them. It is not that we "like" them, it is
that we appreciate them. We can contrast ourselves with them and
decide more clearly what kind of person we wish to become.

But thoughtful citizenship requires more. It requires respect.
Students must feel respected. They need freedom from ridicule and
sanctions in order to feel safe enough to take risks in participating
in class. Bronowski mentioned important values, such as the search
for hidden likeness, the habit of truth, independence of rigorous
thought, and respect for human dignity, and we use them as our
organizing ideas. What does the ethics of respect in science
instruction imply for the development of functional citizenship?

First, we wish to cause students to value respectful dissent. They
must be allowed, even encouraged, to disagree. They must be
respectfully listened to as they disagree with peers and with
textbooks. In fact, they should be allowed the freedom to form
concepts out of their own experiences first, without the implication
that some "authority" should spoon-feed the concepts to them. I
believe that the way typical schools structure exams and give
grades subtly indoctrinates students with a value system which
stifles thoughtful citizenship. If most of what is taught emphasizes
rule following rather than innovative concept building, we should
not be surprised to produce a populace dependent on consensus
reporting and yet suspicious of it.

Our type of instruction is not a permissive environment, nor does it
emphasize disorganized individual science projects. We group
students (16 per class) in groups of like-minded peers. It is the
teacher's job to present a tantalizing question to the group each

day, which can be debated, and concerning which naïve theories may be respectfully revealed.

As a group, students design lab activity which will obtain data concerning the issue. When all students have written the agreed upon lab, they go to lab, working in pairs to collect extensive data. The following day the lab partners report their findings to the class and all the data is often plotted on a class graph. Trends in the data are discussed and sources of error in stray data points are sought with interest. Typically the lab procedure is revised according to the new information gathered from the previous trial.

This is fundamentally different from all other instructional methods. Students' personal data and their individually chosen confidence intervals are recorded on graphs which are laminated and proudly posted at each lab. Each student "owns" their data and their own analysis of it.

Day after day, labs report their data to the class and conclusions are drawn. The teacher guides the discussion to selective comparisons of a variety of previous labs. At times students begin new topics and then, without warning, find relationships to work done several months before. "You set us up!"

With strong guidance of Socratic questioning from the teacher, students are led to formulate their own questions and design their own labs. Students are led to experience the world in such a way that the content upon which they reflect is expanded. Previous findings provide frames of reference through which students view new experiments. Definitions build one upon another. Defining one thing as distinct from another provides fresh focus on the underlying issues. Naïve theories give way to new conceptualizations. But the process of respectful debate and personal responsibility for organizing perceptions of the world continues.

Our future citizens internalize a set of rules, expectations and beliefs about science which encourage creative and ingenious intellectual behavior that come as part of the scholarly content each student is learning. Our students have participated in repeated experiences which prepare them to make new discoveries in art and science and to enjoy the economic growth that brings to society.

This confronts us with the question. What should be the mission of science education in today's society? Should the major focus be a mastery of a new vocabulary? A litany of scientific facts and opinions? A political point of view? Should a textbook be the course outline? Are laboratories for the purpose of verifying what the students have been told or have read in textbooks? Does mastery of words and terms and high scores on question-answer tests indicate a replacement of naïve concepts with deeper understandings?

Will our instructional methods produce a populace who can deal with societal problems creatively? Who can think logically? Who can respond politically to science related societal issues? But, more than basic understanding of science, will our students understand the foundation of a truly humane value system in which there is respect for truth and accuracy and for honestly held differences of opinion?

As science progresses, we honor those whose beliefs we no longer share (Copernicus, Newton) because we honor the process of listening, testing, disagreeing and changing our opinion respectfully.

In *Science and Human Values* (page 67), Bronowski says:

"As a set of discoveries and devices, science has mastered nature; but it has been able to do so only because its values, which derive from its method, have formed those who practice it into a living, stable and incorruptible society. Here is a community where everyone has been free to enter, to speak his mind, to be heard and contradicted; and it has outlasted the empires of Louis XIV and the Kaiser."

"This is a stability which no dogmatic society can have. There is today almost no scientific theory which was held when, say, the Industrial Revolution began about 1760. Most often today's theories flatly contradict those of 1760; many contradict those of 1900. In cosmology, in quantum mechanics, in genetics, in social science, who now holds beliefs that seemed firm sixty years ago? Yet the society of scientists has survived these changes without a revolution, and honors the men whose beliefs it no longer shares.... The whole structure of science has been changed, and no one has been either disgraced or deposed. Through all the changes of science, the society of scientists is flexible and single-minded together, and evolves and rights itself. In the language of science, it is a stable society."

The society of scientists is a stable society. Teaching students this way can make for thoughtful citizenship.

3. Personal Responsibility

We vote taxes for schools. We get government grants for school programs. We talk about administrative structures, elective classes and Charter schools. Wait a minute! Whose job is it to learn in school? The bottom line is, it's the student's. Does the student want to learn? Does the student want to put forth the effort to learn? What does the student think learning is? Memorization? High test scores? Reasoning? Thinking?

The unknown factor in all of this is the attitude of the student. Does the student memorize a conclusion and join an activist group to loudly affirm a point of view? Does the student, through their ages and stages, develop the capacity for rational personalized thought?

There is an inner life of a child. A future citizen. That inner life takes it's reference from society. The sense of self may be just a fragment of group membership. An insignificant part of the whole. Many, perhaps most societies educate children as if they should be "seen but not heard." The child's job is to blend into the group. To perform. To obey. To do their duty. However, with a focus on individuality, the child's sense of self may be an outgrowth of the attention and respect which is given each day. What we do in the family matters. What we do in the classroom matters.

When a young child looks you square in the face and says, "You know, Santa Claus isn't real," you have just witnessed a stride for individuality. The child is considering the available data and suggesting a personal conclusion. That conclusion may falter. It may change with further data. But the right to construct personal conclusions is dawning.

The real life of the child is in the mind. In the relationship of self to events around. Do we, as teachers, notice that? Do we neglect or nourish individual expression. Children want to learn. They look for signals that they are growing up. They have an urge to improve. To engage. To have fresh experiences. But the toxic cloud of endless "facts" they are to memorize and spew back eventually dulls the instinctive curiosity to figure things out.

Learning is personal. Our inner life is personal. Our sense of self within a reference group is individual and personal. This outlook on personal development is central to our country's democratic activity, based as it is on individual rights and personal responsibility. Teachers need to be observant and deeply engaged in nurturing how students see their individuality in a group. To develop appreciation for the process of science, students need to cultivate a sense of personal ownership of their own thinking. We need to notice and celebrate the first tentative steps toward personal responsibility for drawing conclusions. Classrooms dedicated to teaching the process of science should give children opportunities to forge their own connections between data points. Not only will this help future citizens appreciate the uncertainty of scientific "conclusions," but it may help them navigate the challenges of their search for self-identity in their teen years.

We don't just teach facts. We teach a disposition of mind. Thinking for one's self opens the possibility for creativity. Free thinking minds are more able to see connections between seemingly diverse ideas. Students may become aware that they have a cognitive, associative thought train of their own. The ability to think within themselves. It may never be communicated, shared, or explained to others, but it can provide amazing understanding that "I am me."

I give "half – page tests." I ask students to contemplate ideas on the spur of the moment that they have never thought much about. "Why do whales have blubber?" I am asking that they search their memory for any data they may know about the topic and then propose multiple hypotheses. Once in a while a student will write, "I don't know." That tells me that this student has not yet defined themself as a thinking person. This student probably assumes that the job of school is to provide facts to be memorized. Since I have not given the child a fact to memorize, they can't be expected to know it. They give themselves permission "not to know." In order for this child to learn real science, I will have to struggle to nurture their sense of personal ability and responsibility for putting ideas together on their own.

This can start at a very young age. The Reggio Emilia young child education program is based on, *"The principles of respect, responsibility, and community through exploration and discovery."* The child is viewed as being an active constructor of knowledge rather than being seen as the target of instruction. Teachers take notes and create portfolios about the development of reasoning in each child. Much of the instruction at Reggio Emilia schools takes place in the form of projects where they have opportunities to explore, observe, hypothesize, question and discuss to clarify their understanding. Teachers become skilled observers of children's behaviors in order to develop the next appropriate lesson. Teachers document the expressions used by the child and use their drawings to help them "make learning visible". By having their creations publicly displayed, children, at a very young age, develop their sense of individuality and intellectual responsibility.

The atmosphere of the learning environment plays a huge role in developing responsible citizens.

4. A Fundamentally Different View

Many people consider science to be a collection of facts. A group of conclusions that experts have agreed upon. Headlines scream at us to join the consensus. Become activists to support their cause. Their perspective. They know. They know best. Believe what they say. So many bright minds have reached that conclusion that they are obviously, compellingly, right. *"Fat makes you fat." "Eat less fat and focus on fruits and vegetables." "Ninety seven percent of scientists believe in global warming." "Man-made global warming is irreparably harming the earth." "Save the earth."* These are conclusions. Inconvenient truths. We must believe. Scientific authorities tell us so. There is no place for democratic reasoning here. But should we trust them?

Are science and democracy two separate endeavors? Some say science has facts agreed upon by authorities while democracy is reasoned debate. Is that true? People aren't sure. People may not trust atmospheric scientists but they still get on airplanes. When did modern science depart from ancient ways of thinking? What is different, unique about modern science?

From the dawn of humanity, the natural state of man has been to live under tyranny of one form or another. Things have been different for a small portion during the brief few hundred years since the Enlightenment Revolution. Even today, less than 10% of humans alive enjoy individual rights. And these are the cultures that are most prosperous. These cultures produce the overwhelming number of Nobel Laureates. These cultures are not free because they are prosperous but they are prosperous because they are free. Prosperity from scientific advancement. Scientific advancement from the scientific method. The scientific method rests on individual rights and personal responsibility. Individual rights and personal responsibility form the cornerstone of a free democracy.

 Until the Enlightenment Revolution beginning in 1600, Europeans believed in authority. Universities were structured around the ideas of Aristotle. Astronomy was Aristotelian. Medicine was Aristotelian. To diagnose a patient, one consulted the stars. The universe wasn't logical. Greeks told us that air, earth, fire and water had their own personalities and preferred place. They did experiments to notice what was going on but there was no need to make an overarching logical theory because nobody believed there was one. Overly organized churches, governments and universities knew the literature and were powerful enough to impose their ideas on students – and thus on society. By Enlightenment times ideas were changing. After the Dark Ages, a decentralized and fragmented Europe enabled individual organizations. And individual thought.

Observation and experiment became the newly popular method of understanding the laws of nature. Philosophers might have ideas but you needed experimental evidence to make a compelling argument. Acceptable data was collected from information provided by the senses. That meant that ultimate problems could not be solved, so the experiments sought to claim a degree of accuracy and a known margin of error. Variables. Independent, controlled and uncontrolled variables. By 1664 the manner of communicating was formalized in organizations such as the Royal Society of London. They asked for tangible evidence that could be repeated again and again to demonstrate an idea. Just quoting authority was not accepted. Their motto was *"Nullius in verba"* which means to take nobody's word for it. This went beyond simple observation. Now suddenly, investigators were purposefully contriving experiments. Their results were publicly shared and implications were widely debated.

The first published work championing this new world view was published in 1600 by William Gilbert. He was well known, widely respected, and had become the personal physician to Queen Elizabeth. Having been assigned to look after the health of her navy as it prepared for the attack of the Spanish Armada, Gilbert made a habit of visiting ships and sailors. In fact, it seems that he actually talked to them. Real conversations. Interesting conversations. Not the dictates of authority. Gilbert learned about their navigation methods and their puzzlements. He became intrigued with their use of magnetic compasses and set up a laboratory at home to study magnetic effects. He built a spherical lodestone and devised instruments to measure magnetic dip. He designed experiments to see what would happen. This moved reasoning from "natural" science into "experimental" science. But he went further. He concluded that his experiments matched pretty well what the sailors were noticing as they sailed the oceans. Gilbert concluded that the earth itself must be a big magnet. The magnetic poles of earth must be near (but not quite on) earth's rotational poles. This opinion was in contrast to information from prior authorities that compasses were attracted to a northerly star or to a lodestone mountain in the Arctic. Many statements in his published work, *De Magnete*, decry and deride society's mindless quoting of authority. With Gilbert, a new set of values had dawned on society. Individuality. Personal responsibility for actually conducting experiments. A sense of respect and human dignity in reporting findings with the understanding that ideas may change. Our results are dependent upon our experiments using human senses and have a degree of accuracy and a margin of error.

This is the message I want to give my students. Science, as we know it, does not derive from quoting authority. Those *97% of experts* who are quoted in the press are not reporting science. They are harkening back to a way of thinking represented in society before 1600. Scientific thought today is not that at all.

The Enlightenment brought a new mindset to the western world. A mindset of doubting authority and testing ideas for yourself. A mindset of trying new things and observing what happens. This Enlightenment mindset focused on individual invention and sharing of new ideas has led to the remarkable living standard of our country. Not everyone in our country understands. Many citizens mistrust science. They easily revert to authoritarian attitudes and become activists for causes that are "obviously" true. They fail to recognize that the statement, *"97% of scientists believe ... something (anything)"* is an argument relying on pre-Enlightenment logic. They are using the mindset of Aristotle. They are more than 400 years out of date.

And this is not benign. This attitude, when actively pursued by a large enough proportion of citizens, will fundamentally change how society functions. No longer free and growing but authoritarian and brittle.

I struggle and wonder what I can do to help bring our citizens to modern ways of reasoning. Perhaps the best I can do is try to prepare my own students for the citizenship challenges they will find in their future.

5. The Value of Doubting

I have, on the walls of my classroom, quotations by Sir John Cornforth who won the Nobel Prize in Organic Chemistry in 1975. *"Science is the art of the testable. Scientists do not believe. They check."* Cornforth says, *"It may seem odd that a system of knowledge based on doubt could have been the driving force in constructing modern civilization."* Curiosity, skepticism, good communication, and publication of results – whether they agree with your theory or not. That is the essence of research. That is the essence of what we do in science class. Cornforth says real science is different from nearly all other subjects taught. *"Languages, literature, religion, law, art, music, even pure mathematics are all human constructions, and they can be taught on the basis that 'these things are so because men made them so."* … *"The habit of asking questions like, 'Who says so?' 'How do they know?' 'What's missing?' 'What are the assumptions?' 'What is the scale?" Is it all about the same thing?' 'Do the figures make sense?' will make them more receptive to the message of science for the rest of their lives." (RACI 75th Anniversary Lecture, September1992)*

Thoughtful citizens reserve for themselves the right to consider and make their own conclusions.

In *Science and Human Values* (page 61-63) Bronowski says:
"A man must see, do and think things for himself, in the face of those who are sure that they have already been over all that ground. In science, there is no substitute for independence...."

"The profound movements of history have been begun by unconforming men. Dissent is the native activity of the scientist, and it has got him into a good deal of trouble in the last ten years. But if it is cut off, what is left will not be a scientist. And I doubt whether it will be a man. For dissent is also native in any society which is still growing. Has there ever been a society which has died of dissent? Several have died of conformity in our lifetime."

"Dissent is not itself an end; it is the surface mark of a deeper value. Dissent is the mark of freedom, as originality is the mark of independence of mind. And as originality and independence are private needs for the existence of a science, so dissent and freedom are its public needs. No one can be a scientist, even in private, if he does not have independence of observation and of thought. But if in addition science is to become effective in public practice, it must go further; it must protect independence. The safeguards which it must offer are patent: free inquiry, free thought, free speech, tolerance."

"The society of scientists must be a democracy. It can keep alive and grow only by a constant tension between dissent and respect: between independence from the views of others, and tolerance for them."

Tolerance among scientists cannot be based on indifference. It must be based on respect. Respect as a personal value implies, in any society, the public acknowledgements of justice and due honor."

"Science confronts the work of one man with that of another, and grafts each on each; and it cannot survive without justice and honor and respect between man and man. Only by these means can science pursue its steadfast object, to explore truth."..."In societies where these values did not exist, science has had to create them."

In *The Meaning Of It All*, (page 26-27), Richard Feynman points out:

"All scientific knowledge is uncertain."... ..."I believe that to solve any problem that has never been solved before, you have to leave the door to the unknown ajar. You have to permit the possibility that you do not have it exactly right. Otherwise, if you have made up your mind already, you might not solve it." "It is of paramount importance, in order to make progress, that we recognize this ignorance and this doubt. Because we have the doubt, we then propose looking in new directions for new ideas. The rate of the development of science is not the rate at which you make observations alone but, much more important, the rate at which you create new things to test."

"If we were not able or did not desire to look in any new direction, if we did not have a doubt or recognize ignorance, we would not get any new ideas. There would be nothing worth checking, because we would know what is true."

The true essence of science thrives on being doubted. "Never trust a science teacher!" Doubt the teacher and look at the evidence.

6. A Way Of Dealing With It

So how do we help our students find their way in our modern world of conflicting values? On the one hand, science encourages doubt and testing. On the other, news alarmists publicize 97% certainty in reports from scientists. In their focused, individual lab projects my students become comfortable debating variables and posing new lab procedures. Can that help them view the news? Can we make a connection?

Students are young. They don't know how to handle mathematics very well. They can take data and graph it but they are not ready to calculate statistically. A t-test or p-value is meaningless at this age. They are just beginning to control variables and explore alternative hypotheses. They want to embrace scientific doubt but are at an age where they still comfortably follow what they are told in their everyday lives.

It seems to me that one way to lead them along the path of rational discourse as they grow in their maturity is to present Chamberlain's method of *Multiple Working Hypotheses*. Doing so gives them the type of adult authority figure they are still comfortable with who is instructing them how to doubt in an acceptable way.

Chamberlain worked in geology a hundred years ago. No one knew continents moved. Rivers moved. Ice ages came and went. Mountains rose and fell. How could you account for it all? He said that it seemed complex. There was probably more than one cause. Maybe they weren't equal. Maybe one cause was stronger or more immediate, but several causes could be cumulative. He was interested in the Great Lakes basins and tried to figure out how they were formed. He thought there were at least three causes that might have all contributed to the shape of the lakes; crust deformation and mountain building, river erosion and glacial excavation during the last ice age.

In 1890 Chamberlin wrote, *"the dangers of parental affection for a favorite theory can be circumvented"* by looking for more than one reason. Even today, affection or loyalty to a theory may lead news outlets to report evidence to support only the ruling theory, and not consider alternative explanations. Concern for global warming is a prime example. Other examples can be found in ecology, medicine, geology and astronomy. Why did the dinosaurs die? Is the cause of death a heart attack, or is it confounded by diabetes and obesity? Is the expanding universe moving on a "paper folding" paradigm? Complex causes may not have simple answers.

Chamberlain advocates that we look for sequential and simultaneous multiple working hypotheses. We usually can't control all the variables to set up a modern mathematical test. Hooke, Linnaeus, Cuvier, and Darwin did good science before statistical methods were developed. They were able to doubt and debate. They made huge contributions to their fields of study.

In a world where religious and political fundamentalism tells us they have the answers, Chamberlin's worldview that values lateral thinking and multiple possibilities can provide an organizer for the way we teach. What is required is that we *"bring up into view every rational explanation of new phenomena, and to develop every tenable hypothesis respecting their cause and history"* (Chamberlin 1890).

Doubting and debating establish the road to thoughtful future citizens.

7. A Sense Of Urgency

You want the best for your child. But you only get one chance. You have a ten year window of opportunity to lay the groundwork for your child's worldview. After that, teen years take over. What will you do? What experiences can you provide to nurture their understanding?

The scientific method of individual reasoning and personal responsibility is fundamental to our culture. The ideas presented in these essays provide a lens through which to view what is influencing your child's life, Which type of learning is going on in their school? What can you do personally to teach these habits of mind to your children? Ways of reasoning unique to our civilization are precious. We shouldn't be shy about promoting them.

No matter which school your child attends, you can make a difference. I only have a given student for an hour a day. That is enough. They go home with stories to tell. Every day. Science is fun because they are listened to. Because they have the opportunity to be thoughtful. It is not trivial stuff. It is exactly age-appropriate. Reasoning is hard work, but it is very rewarding.

Marian Diamond has devoted her life to studying changes in animal brains (particularly rat brains) when the animal was subjected to varying levels of environmental stimulus. She found that rats exposed to enriched environments grew larger cerebral cortex. This is the part of the brain used for higher level thinking. Rats that could watch other rats play but could not reach the toys themselves had no such brain changes. This suggests our children need active participation in their learning process. TV programs won't do it. Providing children with enriched experiences may have an impact on their cognitive and memory capabilities.

Marian Diamond writes:

"The cerebral cortex, the area associated with higher cognitive processing, is more receptive than other parts of the brain to environmental enrichment. The message is clear: Although the brain possesses a relatively constant macrostructural organization, the ever-changing cerebral cortex, with its complex microarchitecture of unknown potential, is powerfully shaped by experiences before birth, during youth and, in fact, throughout life. It is essential to note that enrichment effects on the brain have consequences on behavior. Parents, educators, policy makers, and individuals can all benefit from such knowledge'

How much is enough? Time matters. Diamond found that as little as an hour a day significantly affected the structure of rat brains.

"The duration of exposure to the enriched environment is clearly a significant dependent variable that must be factored into research in this area. As short a period as 40 minutes of enrichment has been found to produce significant changes in RNA and in the wet weight of cerebral cortical tissue sampled. One day of enrichment was insufficient to produce measurable changes in cortical thickness, whereas four consecutive days of exposure (from 60 to 64 days of age) to an enriched environment did produce significant increases in cortical thickness, but only in the visual association cortex (area 18) (Diamond 1988). When young adult rats were exposed to 30 days of enrichment, however, the entire dorsal cortex, including frontal, parietal and occipital cortices, increased in thickness. Extending the duration of the stay in enriched conditions to 80 days did not produce any greater increase in cortical thickness than that seen at 30 days (in fact, it was often even less); however, the longer the rat remained in the enriched conditions, the longer the cortex retained its increased dimensions following return to the standard environment (Bennett et al, 1974). When we looked at age-related differences in the context of duration of stay in the enriched environment, we found that old rats (766 days of age) placed in enriched conditions for 138 days showed an increase in cortical thickness that was quite similar to that observed in young adult rats (60 days of age) that had lived in enriched conditions for 30 days.
Response of the Brain to Enrichment
by Marian Cleeves Diamond
http://education.jhu.edu/PD/newhorizons/Neurosciences/articles/Response%20o
f%20the%20Brain%20to%20Enrichment/

Until recently, neuroscientists believed that once the brain completes its development, it is unable to change. Now we know differently. What can we do?

Well, listen and read to them. Pose interesting things to discuss with them. Contemplate" what if' and "how do you know?"

What if:
- The earth is not flat.
- The sun does not move across the sky each day.
- Our school hill is not 50° .
- Sunlight does not cover everything.
- The earth had no moon.
- Continents are not this way forever.
- Magnetic North is not a big iron mountain up there somewhere.
- Airplanes do not sit out on the tarmac to annoy us.

Look at all available data whether you agree with it or not.
Search for uncontrolled variables.
Propose multiple working hypotheses.
Arrive at a tentative conclusion.
Don't expect them to agree with you or with each other.

Dinner time may be a great time to pose such conversations. You are gathered around a table anyway. You might as well talk about something interesting.

Of course, there are many other opportunities. Take your child shopping to a hardware store or a fabric store. Talk about what you see and theorize about what things could be used for. Purchase a few things to play with. Follow up with little activities to test and explore ideas.

But if you send your children to our school, perhaps they will be in my class. Remember that I am trying to cause them to discuss ideas each day. To think as individuals. To embrace personal responsibility for their own thinking. To learn some science concepts along the way. To prepare for a future as thoughtful citizens.

And we do funny things in class that mark a memory. Things they come home and talk about. And no matter what they tell you, remember to Never Trust A Science Teach

8. Students said:

Thanks for an awesome year in science. I love learning about the heart, Happy Valentine's Day.
ML

Dear Doc "O"
Thank you for being an awesome teacher! You have taught me so much about the heart, tall ships and whale blubber.
P.S. I don't know how many calories are in this chocolate bar. It says 40 calories though.
DK

I really enjoyed all the labs in science. I think science is one of my favorite classes because we get to find things out for ourselves. We don't have you tell us everything.
KY

In science I like the way you let us try to find out the answer without telling us.
AD

I like that we can be individuals and responsible for ourselves.
NB

Dear Doc "O"
Thank you for being a great teacher. What makes me come to school is that you let us choose in a safe way. You don't just tell us how to do something. I like how you teach us how to do something and let us do it and not monitor us.
ZW

I love science because it is not all paper work. I like physically experiencing what we are learning and having us figure things out. I also like doing dangerous things safely to make it more fun. Thanks for being an awesome teacher, Doc "O".
NB

Thank you for teaching me how to do a lab and being amazing, awesome and the best teacher ever!!
AS

I like the fact that instead of you telling us about science stuff, we actually do it.
GN

I like that we actually do labs and do them in a fun way. I like that we figure out our data by doing the tests ourselves. I also like how you teach.
MF

I like science because you get to do once in a lifetime opportunities and debate about things you don't think you are going to debate about. For example, Spider man, doughnuts. I like that you think outside the box.
AH

Science class is different. It's not like the other classes we have. It's hard to miss class because it goes on without us. Plus it is really fun.
LH

Science class isn't just fun, it is interesting. You always get to do something new. People are always saying, "I don't want to miss school because of science class." They say this because they know they are going to do something cool and if they miss it – they miss it. The airplane takes off. It's gone.
WL

I like science class because we always do something fun, even if we write practically the whole time. Doc "O" never teaches us boring subjects, or at least teaches in a way that makes it incredibly interesting. I never want to miss science because, as Doc "O" says, "The airplane takes off." Even if it wasn't the best class, it's always important the next day. Doc "O"'s lessons always sink in. She makes everything amazing. I am disappointed that I only have her for two years.
MF

I like science because we always do something special. We never expect something. We have a lot of responsibilities.
MP

It is the way you teach and everybody thinks you are the best teacher ever!!!
BT

I love science because I get to play around with stuff that is not always straight taught as you-have-to-do-this class. Science is the reason school is terrific for me. I love having control over fire. Thank you Doc "O" for making science the best class ever.
 GH

I love coming to science class because every day is something new. The units are very interesting, and I like the fact that you let us figure out everything by ourselves. I really enjoy the way this class is taught.
CL

I like science class because we get to explore all kinds of things! One of my favorites was the poison unit. I like the poison unit because there was a lot of being very precise.
ML

I like science class because you can look at things a new way and use things differently, such as a j-tube becoming a keel.
MM

I like science class because of how intense it is, and when I am absent or have to leave early, I always wonder that we did in science today and feel bad. My favorite units were the Engineering Event and the Hot Rod Races unit. Science is one of my favorite subjects at school!
SM

I like science because it's interesting. Also it's captivating the way you teach. I don't want to miss science class because you feel really behind. It is really fun doing lab!
EL

Thank you for making science really interesting, different, and confusing.
ZW

Thank you for making science class different. Thank you for being confusing. Thank you for making class fun. Thank you for making it challenging.
CN

I like that you give us a lot of freedom. For example, you weren't just staring at us when we measured and weighed poison.
JP

I really like how we get to do stuff that is somewhat dangerous, unlike last year. I just like that satisfying feeling of striking a match or solving a question.
CP

Thank you for teaching me how to do a lab and being amazing, awesome and the best teacher ever!!   AS

About the Author   Meredith Olson Ph.D.

Dr. Meredith Olson, known affectionately as Doc "O" to her students, has taught elementary, middle school and high school math and science in Seattle for nearly 60 years. Her primary goal is in improvement of pre-college engineering education. By going to lab to work on contraptions every day, her students come to understand properties of the mechanical world.

"It has been a long and interesting trip. Studying some metallurgy in grad school. Evening classes. After a full day of high school teaching. Consulting for JPL as the Mars Pathfinder Educator. Weekends. Working in the summer with UNESCO in Zimbabwe, Kenya, and Uganda. Teaching dozens of weekend and week-long summer teacher workshops in South Carolina and Montana. Being a consultant and curriculum designer for Health and Physiology education in Washington, Oregon, Idaho, Montana, and Alaska. Being a summer adjunct University instructor for more than 20 years in Seattle, Idaho and Montana. Teaching teachers. Teaching students every day, every year for 59 years. Observing how learning happens. Becoming aware when real learning isn't happening. When it is just "show." When it is just teacher–pleasing to get a grade. To get a credit. To get a university degree."

See Dr. Olson's open letter outlining her philosophy of lesson design, available on the JPL website - Exploring Preface pp 11-13 http://mars.jpl.nasa.gov/education/modules/GS/GS07-19_preface.pdf

Dr. Olson believes that children must construct their own understanding from active design and assemblage of contraptions. By testing, failing, remodeling, and trying again, we come to see the structure when we look. By carefully examining materials we have, we may perceive how to use them in new and unexpected ways. Children begin to understand the engineering process. Besides, it is fun!

www.ingramcontent.com/pod-product-compliance
Lightning Source LLC
Chambersburg PA
CBHW061759040426
42447CB00011B/2375